# TOUGH MEN IN HARD PLACES

## A PHOTOGRAPHIC COLLECTION

from the Western Colorado Power Company Collection,

Center of Southwest Studies, Durango, Colorado

Esther Greenfield

WESTWINDS
PRESS®

Greenfield, Esther.
  Tough men in hard places : a photographic collection /
Esther Greenfield.
     pages cm
 Includes bibliographical references.
 ISBN 978-1-941821-12-1 (pbk.)
 ISBN 978-1-941821-49-7 (hardbound)
 ISBN 978-1-941821-33-6 (e-book)
1.  Rural electrification—Colorado—History. 2.  Electric
utilities—Colorado—History. 3.  Electric industry workers—
Colorado—History.  I. Title.
 HD9688.U53C657 2014
 333.793'209788091734—dc23
                                        2014017813

Design by Vicki Knapton

Front cover: Wear and tear on Trout Lake flume,
Telluride, CO, May 1921. *Photographer: P. C. Schools.*
Back Cover: Constructing dam, Ouray, CO, c. 1910.
*Photographer: Chase & Lute, Ouray, CO.*

Published by WestWinds Press®
An imprint of

GRAPHIC ARTS
BOOKS®

P.O. Box 56118
Portland, Oregon 97238-6118
503-254-5591
www.graphicartsbooks.com

*For Samuel and Judith Greenfield.*

Ames Power Plant, Ames, CO, c. 1910. *Photographer: Unknown.*

Raising poles near Tacoma Power Plant, Durango, CO, August 1930.

*Photographer: P. C. Schools.*

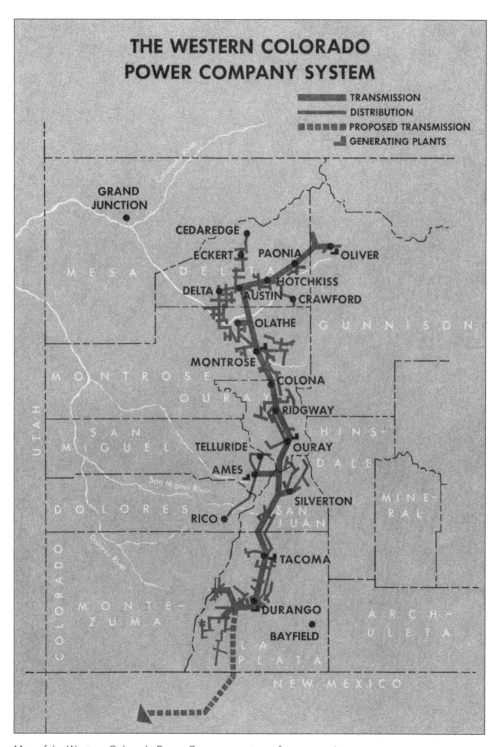

Map of the Western Colorado Power Company system of power stations. *Artist: Unknown.*

Riveting pipe, Cascade flume, Durango, CO, 1924.
*Photographer: P. C. Schools.*

# FOREWORD

Esther Greenfield's *Tough Men in Hard Places* brings us the story of industrial development in the San Juan Mountains of Western Colorado during a thrilling period of western American history. This book shines a light on the unique story of electrification and the stabilization of electric power supply systems on Colorado's western slope. The man at the center of the story, P. C. Schools, was and remains one of the fascinating personalities of the development of communities focused on mining, ranching, and later, tourism in this most beautiful section of Colorado.

This book draws from an increasingly useful collection of documents and photographs deeded to the Center of Southwest Studies from the successors of the Western Colorado Power Company. That company, P. C. Schools's employer, is an example of the new conglomerations of energy production businesses in remote regions in the late nineteenth and early twentieth centuries, a period synonymous with the Progressive Era and a second, massive wave of industrialization in America and its western states. The collection is a source of pride for the Center and is growing in popularity with researchers interested in the growth of the American electrical grid and a renewed focus on the historical environmental conditions of power production in the United States. As this book demonstrates, the collection holds many treasures for those seeking the history behind the growth of industrial and consumer power production in this region. A better case study would be difficult to find.

Esther focuses attention on the processes of modern development in the San Juan Mountains region. She is one of the dedicated volunteer researchers whose work expands access to the archival offerings of the Center of Southwest Studies in its Delaney Southwest Research Library and Archives. For her efforts and this book the Center is grateful. We invite the reader of this volume to consider the unique situation of the electrification projects of the West through the experience of power company workers and supervisors of the early days in Colorado when water flowed down mountainside flumes to power lights in mills and mines, cabins and saloons, and when a single snow slide could halt all power distribution in a few frightening moments. Here is a story of American progress in an emerging industrial area where the deserts of the Southwest meet the Rocky Mountains.

Jay T. Harrison, Ph.D.
Director, Center of Southwest Studies
Fort Lewis College
Durango, Colorado

# INTRODUCTION

In the late 1880s, the once-booming gold and silver mines in southwestern Colorado were in deep trouble. Their rich veins of ore were vanishing like wispy dreams and operating costs were soaring. To cut costs and generate cheaper power, mine owners built ore processing mills next to the mines and powered them with steam engines fueled by ever-decreasing supplies of wood and coal. Although they managed to produce an adequate supply of electric power this way, it was very expensive. More importantly, the power generated, which was direct current power, could not be sent over the long distances required to reach the mines scattered around the rugged mountains. As fuel and transportation costs continued to rise, mine owners retreated into bankruptcy and watched their profits go up in coal smoke (*The First 50 Years*, 1963).

The spark of an idea that would solve their problems came, at least on the surface, from the unlikeliest of men, a slight, oddly shaped young man named Lucien Lucius Nunn. He was an Ohio farm boy-turned-carpenter-turned restaurateur-turned-lawyer who made the brilliant breakthrough of imagination necessary to bring affordable electricity to rural Colorado. Here is how he did it.

During the 1880s and 1890s, Nunn was in the "center of a scientific and educational whirlwind that swept its way across Colorado, the United States, and the world" (Buys 1986).

Nunn, who managed the financially faltering Gold King Mine near Ames, Colorado, had ideas about how to deliver cheap power to the mines—ideas based on his knowledge of Thomas Edison's commercial inventions with electricity, George Westinghouse's experiments with alternating current, and his understanding of the alternating current generator that scientist Nikola Tesla had developed. It was said of Tesla that "abstract concepts of energy flashed in his head like bolts of lightning" (Buys 1986) and that he gave little thought to practicalities. But Nunn didn't have the luxury of being a pure theorist or scientist. His ideas had to bridge those of Tesla, Westinghouse, and Edison and be practical. And, more importantly, he was in a hurry to find a solution to the mining industry's problem or he and his wealthy backers back east would be out of a job.

Desperate to turn things around quickly, Nunn recruited a team of what he called "pinheads," smart, risk-taking young engineers and scholars, and formed a think tank whose task was to find innovative solutions to the problem of building a generating station that could transport electricity over long distances and in particular to the Gold King Mine at Ames, Colorado, using the new phenomenon of alternating current electricity. It would not be easy. His technology would have to work despite icy temperatures, blizzards, avalanches, and the wild

The world's first hydroelectric plant, Ames, CO, 1890. *Artist: Unknown.*

summer lightning storms so common to the San Juan Mountains.

During the winter of 1890, Nunn's team built a "crude" wooden shack at Ames, Colorado, in which they installed the necessary machinery, including the experimental generator designed by Tesla. Then the men strung copper transmission lines, which sparkled and glowed golden in the sun as they swooped the nearly three miles from the newly built Ames Plant to the Gold King Mine.

Finally, in the spring of 1891 all was ready to go. Would it work? No one knew for sure. Nunn threw the switch and a "brilliant arc of electricity shot six feet into the air" (Buys 1988). The Gold King Mine was electrified. It was a monumental event that produced economical power and it transformed the commercial application of electricity forever.

Thereafter, according to eyewitness reports at the time, the generator ran smoothly and cheaply and provided great entertainment to the local communities. "Large crowds were apt to gather on Sunday evenings to watch the spectacular start up of the power plant's machinery. It seemed incredible to them that electrical current—or anything—could travel 186,000 miles per second."(ELLISON)

## POWER COMPANIES EMERGE

After that "brilliant arc of electricity shot six feet into the air," power

**E. J. Braund, court reporter for the 7th Judicial District:**

**"I remember the first electric lights. Customers had to light a match to see if the lights were burning. Everyone kept gas or oil lamps handy for use after midnight when the electric plant was turned off."**

*Interview, August 1939*

companies sprang up like wild mushrooms after a summer rain. The companies were eager to capitalize on this new opportunity, which they suspected would be more lucrative than all the gold and silver booms of the past. At one point there were about thirty separate power companies operating in Colorado to bring electricity, power, and conveniences to rural homesteads. One by one, however, they went out of business or merged. In March 1913, the remaining companies were consolidated into the Western Colorado Power Company (WCPC) and it was their donation of thousands of photographs to the Center of Southwest Studies that formed the basis for the collection of photographs in this book.

From interviews recorded with farmers, ranchers, and former employees of the WCPC thirty or more years after they retired, what follows are firsthand descriptions of those early days, when electricity first came to their homes and businesses.

Ames Power Plant, Ames, CO, c. 1891. *Photographer: Unknown.*

Cutting through solid rock near Tacoma Plant, Durango, CO, 1928. *Photographer: P. C. Schools.*

## TOUGH MEN IN HARD PLACES

Crouched down behind a tumble of boulders on the mountain pass, the lineman tried to stay out of the wind as he got his tools out of his leather bag. He could smell another snowstorm coming and he and his crewmates wanted to finish their line before snowshoeing back to the boardinghouse and the warm meal waiting for them there. The air was thin at these elevations—between eight thousand and thirteen thousand feet—but these were rugged men who worked despite the handicaps of weather and rough terrain of the Colorado mountains. Blizzards, swift and sudden snowslides, deadly lightning, and mudslides were everyday occurrences for them. Summer and winter, rain, snow, and rockslides didn't stop most of them from doing their job. Sometimes, however, it all just proved to be too much for even the toughest of the men. Despite the good pay, some just walked away and disappeared, leading to turnover that was at times as high as 75 percent.

## LOOKING FOR WORK

Considering the economic realities of the times and the difficult working conditions, the pay for the men working for the power companies was considered good. For example, in the early days a teamster could expect to get $60 per month, a patrolman $90 per month, a switchman $15 per month, and a watchman $10 per month. A housekeeper or cook at one of the power stations or boarding-houses earned $75 per month. A lineman $124 per month. The interestingly titled "trouble-man" was paid $40 per month and electricians were paid $40 per month. Hourly laborers, hired locally wherever the job was, were paid about 50 cents per hour. Pay was docked a bit for commissary supplies and later for dues to the International Brotherhood of Electrical Workers Union. The men typically worked one of three shifts: 8 A.M. to 4 P.M., midnight to 8 A.M., or 4 P.M. to midnight.

---

**Nate Shute, former WCPC employee:**

**"I worked for the Animas Power Co. for two years while we built the Tacoma Dam. Worked for Skidmore and Loftus who had the contract to clear all the trees and brush for fifty feet around the dam site. I got paid $1.50 a day for clearing the trees and peeling logs."**

*Interview, September 1939*

---

## DEATH AND DISASTERS

The crews suffered numerous accidents, most involving strained backs, abrasions, and "foreign objects" in the eye. But, despite their best efforts at safety, there were also more grievous injuries. Men were electrocuted, their bodies were crushed, they suffocated under snowslides, and drowned in flumes in dispiriting numbers.

P. C. Schools with hammer, loading transformer for Durango, November 1909. *Photographer: Unknown.*

Closing old gates at Cascade Dam, Durango, CO, 1909.
*Photographer: P. C. Schools.*

Repairing leaking dam, Ouray, CO, 1931.

*Photographer: P. C. Schools.*

Lineman, Ophir Pass, Silverton, CO, c. 1917.

*Photographer: P. C. Schools.*

**". . . The men were stronger and worked better. They were tougher and could stand more work. Now the boys have to have pajamas to wear."**

Lime Creek Camp, Durango, CO, June 1930. *Photographer: P. C. Schools.*

## Wm. H. Compton,
former WCPC employee:

"I went to work digging holes on the line being run to the Mountain Queen mine and strung wire from the Gold Prince mine over hill to the Sunnyside mine. It always snowed when we strung wire. I worked moving transformers; eight horses and eight men. It took thirty days. I used to set poles—one mile took six men and one team ten days. The men worked nine hours a day.

"Summer work crews usually had a cook with them, but in the winter the men usually could find board at the many small mines all over the hills. Snows were much more and deeper before 1923. After that they were less. We all worked with snowshoes carrying everything on our backs or on toboggans. I broke my leg once when the snow began to run.

"Troubleshooting on the lines was a great problem— always the supplies that had been cached out were stolen— wood stoves, axes, bedding. Chittick and I would go out with packhorses, grub, and a bedroll, working on the poles and staying out till the work was done. We didn't have liquor on the job, but plenty of liquor off the job. The living was high, wide, and handsome, with gambling, dance halls, and what you desire always available. In the early days there was a better spirit in getting work done. The men were stronger and worked better. They were tougher and could stand more work. Now the boys have to have pajamas to wear."

*Interview, September 1939*

Hauling pipe, Cascade flume, Durango, CO, June 1923. *Photographer: Unknown.*

Mr. Eubank shows how he slipped and fell into flume, near Telluride, CO, 1931.
*Photographer: TJR.*

Installing dead-end clamps, Rockwood switch tower, March 1929.
*Photographer: Unknown.*

Building Ilium flume, Telluride, CO, 1923. *Photographer: Unknown.*

Walking the railroad tracks, Silverton, CO, c. 1910.

*Photographer: P. C. Schools.*

Howard's Fork pipeline, near Telluride, CO, August 1942. *Photographer: P. C. Schools.*

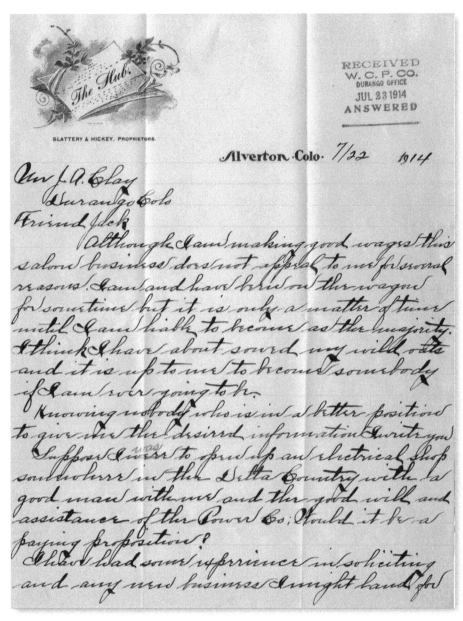

The Hub.

SLATTERY & HICKEY, PROPRIETORS.

Silverton Colo 7/22 1914

Mr J. A. Clay
Durango Colo

Friend Jack

Although I am making good wages this saloon business does not appeal to me for several reasons. I am and have been on the wagon for sometime but it is only a matter of time until I am liable to become as the majority. I think I have about sowed my wild oats and it is up to me to become somebody if I am ever going to be.

Knowing nobody who is in a better position to give me the desired information I write you

Suppose I were to open up an electrical shop somewhere in the Delta Country with a good man with me and the good will and assistance of the Power Co, Would it be a paying proposition?

I have had some experience in soliciting and any new business I might land for

"Although I am making good wages the saloon business does not appeal to me. . . . I am on the wagon for some time but it is only a matter of time until I am liable to become as the majority. I have about sowed my wild oats and it is up to me to become somebody if I am ever going to be. . . . I could do wiring and furnish electrical fixtures. . . . let me hear from you at your earliest convenience."

*James C. Cole, Silverton, CO, July 22, 1914*

Wreck of Big Black Slide trestle, Ilium flume, Telluride, CO, c. 1915. *Photographer: P. C. Schools.*

## Randall Priest,
former WCPC employee:

"I remember they started building the Trout Lake Dam late in the summer of 1895 and it was finished in February 1896. They did the grading mostly with teams of horses and scrapers. There were two construction camps in operation, one on Butterfly Flats and one near the upper end of the flume. There was a big Swede working on the Lake Hope Tunnel who made "big" wages—he got 10 cents per pound for packing from Trout Lake to Lake Hope."

*Interview, 1939*

Inside Lake Hope tunnel, Telluride, CO, c. 1910. *Photographer: Unknown.*

Working on Ilium flume, Telluride, CO, 1923. *Photographer: Unknown.*

Lunch break for Bob Yeager's "bull gang," Tacoma Power Plant, Durango, CO, June 1945. *Photographer: P. C. Schools.*

Laying pipe, Howard's Fork, near Telluride, CO, 1942.
*Photographer: P. C. Schools.*

Constructing dam, Ouray, CO, c. 1910. *Photographer: Chase & Lute, Ouray, CO.*

Power lines in alley, Ridgway, CO, May 1941. *Photographer: P. C. Schools.*

## Ed Duke,
former WCPC employee:

"I remember once when the stockholders of the old Hotch-kiss Packing & Power Co., being dissatisfied with the business methods used in operating the plant, asked for an audit of the books. The owner objected to an audit and his wife sat on the company books to prevent the auditors from seeing them."

*Interview, October 1939*

## Frank Turner,
Montrose, CO:

"About 1905 I was living at 214 S. Third Street here. The Montrose Light Co. set a pole in my alley and set it too close to my barn and too far out in the alley. I asked them to move it. They promised to move it but never did, so I climbed up onto the barn, took a pole, and broke the insulators and pushed the pole over.  I left it right where it fell and the company reset the pole, this time in the right place."

*Interview, August 1939*

Two men on top of Heine boilers, Durango, CO, November 1917.
*Photographer: P. C. Schools.*

Disastrous fire, Gold King Mine, transfer station, Ames, CO, June 1908.
*Photographer: Unknown.*

Interior, Sunnyside Mine, Silverton, CO, June 1910. *Photographer: P. C. Schools.*

"Electric and power brings happiness into your home," Montrose, CO, c. 1920. *Photographer: Orr Photo.*

**Enoch J. Shepherd,**
former WCPC employee:

"I worked in my brother's blacksmith shop at Ridgway for about ten years and often sharpened tools for employees of the Light Company. I remember when my family first came to Canon City in 1888 we used candles that we made ourselves as well as the light from the fireplace to light our house. Then someone discovered an oil spring in the hills nearby and we went there to dip oil out in buckets. Some we used for lighting our houses and some was taken to Denver and traded for whiskey. Another method for lighting our house was to melt lard in a pan and tie a three-cent piece to a rag for a wick."

*Interview, August 1939*

Grinding and scraping rust off runner and polishing, Ames Power Plant, Ames, CO, 1923. *Photographer: P. C. Schools.*

Silverton Substation, July 1939. *Photographer: Unknown.*

T. W. Monell,
postmaster, Montrose, CO:

"I remember the pioneering days of the light and power companies in this territories [sic]. The plants were crude but considering the difficulties the service was good. But the lights were
out often and the first electric lights looked like a bunch
of fireflies."

Motor-operated needle drive, Tacoma Power Plant,
Durango, CO, 1937. *Photographer: P. C. Schools.*

Durango steam plant "thaw wagon," 1891. *Photographer: Richard Beatty.*

Power lines, Silverton, CO, 1929. *Photographer: P. C. Schools.*

Ore sample tables, Tacoma Power Plant, Durango, CO, 1906. *Photographer: Unknown.*

## E. E. Hufty,
former WCPC employee:

"As Paonia grew, Beezley's power plant was too small to supply the town with electricity. So a group of us fellows loaned the town $10,000 for a new plant, which became the Oliver Power Co. My light bill was $12 in those early years. In 1936 under the WCPC I used twice as many kilowatts and my bill was only $6."

*Interview, August 1939*

Oliver Plant, Gunnison, CO, July 1939. *Photographer: P. C. Schools.*

## Thomas Mowatt,
Montrose, CO:

"I had the first electric iron—a tailor iron— in town. Ouray owes much to the Western Colorado Power Company."

*Interview, September 1939*

## J. L. Kemp,
WCPC employee:

"In 1926 electric rates were $1.20 a month for lights. Heat was $2.10 a month."

*Interview, August 1939*

Telluride citizens thrilled to see Christmas tree lit with electric lights, 1916. *Photographer: P. C. Schools.*

"When the lights would dim customers would call up the plant and say 'Throw on another shovel-full of apple peelings!'"

Tacoma Plant, Durango, CO, 1905. *Photographer: Unknown.*

**P. W. Wood,** former WCPC employee:

"I was the manager of the Hotchkiss Packing & Power Co. around 1909. Electric service in those days was very poor, the voltage was low, and people in town were put to much inconvenience. Many people claimed the furnace was fired with dry apple peelings, which didn't give enough heat to keep up a full head of steam. When the lights would dim customers would call up the plant and say 'Throw on another shovel-full of apple peelings!'

"In the fall of 1928 the company started twenty-four-hour service. Previous, service had been only until midnight and only began again at 10 a.m. on Mondays and noon on Tuesdays so housewives could do their laundry. Rate of service was fifteen cents per kilowatt."

*Interview, October 1939*

**DURANGO
HERALD-DEMOCRAT**

# Merle Swire Drowns in Flume

Body Found In Pond Below Cascade
Flume After Riding Nearly Mile In
Three Feet of Fast and Cold Water

*Headline*, Durango Herald-Democrat, *May 29, 1939*

Site where Merl Swire drowned, Cascade flume, Durango, CO, 1939. *Photographer: P. C. Schools.*

## J. A. Bullock,
former supervisor at WCPC:

"The Telluride offices were well equipped with rugs, desks, chairs, etc. and there were three bedrooms and a bath for use by employees not having a home of their own. There were usually three to five engineers there, most of whom were young engineering school graduates. We had a stable with good saddle horses. We billed customers on a strictly demand basis with rates varying from $4.50 per h.p. [horsepower] to $10 per h.p.

"I remember in 1909 the Trout Lake Dam broke due to excessive rains and caused untold amount of damage to us, the railroad, and the county roads. Fortunately no lives were lost. It was difficult to find enough men to carry out the repairs and I had to go down there personally to hire farmers with teams of horses.

"We had trouble all the time with the Ilium Flume due to the hillside slipping and taking the flume out numerous times.

"In 1908 we decided to establish a service and repair department due to demands and we went to considerable expense to fit up a suitable shop and to provide special tools and equipment. But it never proved to be a success and the shop idea was abandoned.

"L. L. Nunn established the Telluride Institute so he could train young men for electrical work. He gave them the opportunity to study and learn the theoretical as well as the practical. They hired a college graduate to give the boys operating at the plants regular instruction and I doubt not that numerous boys secured an education in this way that otherwise would not have been possible. Mr. Nunn was beyond a doubt the real Pioneer of long distance transmission of power."

*Letter dated October 1939*

Watchman's quarters, Tacoma, CO, 1905. *Photographer: Unknown.*

Savage Basin junction house, Telluride, CO, June 1916. *Photographer: P. C. Schools.*

Superintendent's office, Ames Plant, Ames, CO, 1915. *Photographer: Unknown.*

Generator, Tacoma, Durango, CO, c. 1910. *Photographer: Unknown.*

Gate and needle valve, Sunnyside Mine, Silverton, CO, 1911. *Photographer: P. C. Schools.*

Mr. Clellan of New York office atop the world, 13,250 feet, Telluride, CO, 1913. *Photographer: Unknown.*

Ptarmigan Lake crew, near Buena Vista, CO, 1913. *Photographer: P. C. Schools.*

## W. F. Wilcox,
justice of the peace, Montrose, CO:

"On the Sunday before Labor Day the dam gave way and flooded the valley doing much damage to the power company. I drove down and found the power plant filled with mud. I think we got light from electricity brought over the mountains from Ouray. The RR, telegraph, and telephone lines were washed out and no mail was delivered in Telluride for weeks. The first pack train through carrying supplies to Telluride was loaded with barrels of beer. It seems the slogan of the times should have been 'No matter what happens, the beer must go through!'"

*Interview, August 1939*

J. A. Clay at unfinished pipeline joining Forebay Lake with Tacoma Power Plant, September 1906.

*Photographer: P. C. Schools.*

Flood, Cascade Creek, Durango, CO, 1927. *Photographer: P. C. Schools.*

Flume damaged by ice, Cascade, Durango, CO, 1927. *Photographer: P. C. Schools.*

Washout of Howard's Fork trestle, Telluride, CO, 1927. *Photographer: P. C. Schools.*

Ice-buckled Ouray pipeline and bridge, 1938. *Photographer: P. C. Schools.*

## From *History of Cedaredge Light Franchises:*

"Starting around 1910 there were several requests for franchises for light and power companies in Cedaredge, to be financed by men in Denver. There were lots of big plans but for some reason they failed to complete them. Finally in 1921 Fred and Warren Parker got the idea to install a plant in Cedaredge. It looked like a discouraging venture as it seemed the town was too small to support a plant of this kind. But they weren't daunted and started the plant. The first boiler was bought in Paonia when the canning factory there was dismantled. During the time of service the plant was down only for twelve minutes on account of trouble. This is a very good record. At first it operated only from sundown to 1 A.M. From I A.M. to sunup the town was in darkness. In 1929 the Cedaredge Electric Plant passed into the hands of the Western Colorado Power Co."

*Undated, presumably 1939*

Mayor and city councilmen see that a fire stream directed against live wire is harmless, Silverton, CO, 1930. *Photographer: P. C. Schools.*

Rewinding generator, Ames Plant, Ames, CO, 1917. *Photographer: Elmer Evans.*

Boring out buckets at Ames Power Plant, Ames, CO, c. 1917. *Photographer: P. C. Schools.*

## Fred Baxter,
former miner in Telluride:

"I worked at the Smuggler-Union Mine and remember when the Liberty Bell snowslide killed three or four men. The rescue party that went to dig out their bodies was struck by a second slide and twice as many men were killed. I often had to repair the tram towers that had been wrecked by slides and it was a scary job. One man would work on the tram while twenty other men watched the mountain above looking for snowslides."

*Interview, August 1939*

Snowslide buried buildings and killed Chap Woods, Camp Bird Mine,
Telluride, CO, 1936. *Photographer: Unknown*

Severe wind- and snowstorm, Animas Forks, Silverton, CO, elevation 12,850 feet, October 1929. *Photographer: P. C. Schools.*

**John Roper,** former WCPC employee:

"In 1885 I operated the San Antonio Mine in Champion Gulch near Silverton. During the winter of that year, I barely escaped being in a snowslide that destroyed the Dutton Mine where nine men were killed. A force of ninety men dug out the corpses and pulled them to Ouray in snow up to their necks.

"I was caught in a snowslide near the Yankee Girl Mine near Ouray. My life was spared but my friend Paul Seebach was killed."

*Interview, August 1939*

Snowslide-damaged flume, Trout Lake, Telluride, CO, 1943. *Photographer: P. C. Schools.*

Fire, Ilium Power Plant, Telluride, CO, 1923. *Photographer: Unknown.*

Patrolman on skis in deep snow at Savage Basin,
Telluride, CO, c. 1930. *Photographer: P. C. Schools.*

Remains of lightning arrester after fire, Savage Basin, Telluride, CO, 1944. *Photographer: P. C. Schools.*

# IMMEDIATE REPORT OF ACCIDENT.

Employee of __The Animas Power & Water Company,__

_____St.; City,___ __Silverton__ State, __Colo.__

Date and hour of accident,_____ __July,3,__ 190 __7,__ __3.50 P.M.__

Date of this report,_____ __July,5,__ 190 __7.__

| | |
|---|---|
| **INJURED PERSON.** | Name, __R. C. Lockwood__ Address, __Silverton, Colo__<br><br>Age, __23__ . Occupation, __Wireman__<br><br>Weekly wages, $ __22.75__ . Married or single? __Single.__<br><br>General duties? __Wiring.__ |
| **The machine, appliance, or thing immediately causing the accident.** | What was it? __Can of gasolene.__<br><br>In whose control at the time? __Nobody.__<br><br>Was it sound and in good working order? _ |
| **CONTRIBUTING CAUSES.** | Carelessness of injured person? __Yes.__ Violation of rules? _<br><br>Negligence of a fellow workman? __Yes.__ |
| **THE ACCIDENT.** | Place, __Substation of Animas Power & Water Co.__<br><br>Description, __Lockwood was lighting a gasolene torch and flames from torch lighted the wood on can of gasolene. In carrying can out of station, his hand was burnt.__<br><br>Names and addresses of witnesses: __NO witnesses.__ |
| **THE INJURY.** | Nature and extent, __Hand burnt. (Not serious.)__<br><br>Was surgical aid rendered? __Yes.__ When? __Shortly after accident.__<br>By whom? __Drs. Fox & Wiser.__ Where? __Silverton__<br><br>Received by employer, __July, 5, 1907.__ 190___ .<br><br>767. 2-1-'06.  Employer. |

Accident report concerning burns suffered by R. C. Lockwood, wireman, 1907.

### Edgar Williams, former WCPC employee:

"I was at the Liberty Bell Mine in Telluride in 1898 when ten snowslides ran destroying hundreds of yards of power lines. It took three months to rebuild the lines because the snow was so deep and the terrible working conditions."

*Interview, September 1939*

### Walter Wright, electrician for the WCPC:

"Around 1901 I was working at the Revenue Tunnel when I saw a snowslide run down the mountain and take out about 500 feet of power line. It carried away eight or nine poles and about 1,500 feet of wire."

*Interview, August 1939*

### William Rathmell, former relief operator, WCPC:

"Once during my shift a snowslide swept out the line between Ouray and Telluride. The broken power line fell across the company telephone line and the plant telephone was burned to a crisp. I remember a flood came down Oak Creek Canyon and I went into the plant in water hip-deep to shut off the plant."

*Interview, September 1939*

Ice damage, Ouray pipeline, 1938. *Photographer: P. C. Schools.*

Silverton district "dog team" hauling toboggan loaded with material to repair a line break, 1910. *Photographer: P. C. Schools.*

Donald O'Rourke,
WCPC office, Telluride, CO, 1952:

"Some of the most rugged country in America is patrolled by employees of the Western Colorado Power Company. Once, Forest White and Harry Wright, patrolmen from the Telluride Plant, were inspecting lines at 13,200 feet. They were on snowshoes when all at once the snow broke loose above them and 'down the mountain' they went. They rode on top of the slide for a while when they fell and went under the snow. They stopped 600 feet from where they'd started. Harry was buried up to his armpits but had his head out of the snow. Forest was thrown clear of the slide just as it stopped. He managed to dig Harry out, they found their snowshoes, and continued their patrol. Perfect service from the power company may not mean much to a customer sitting comfortably in warm homes and offices, but what tales of courage might be told of that same service behind the scenes!"

Road to Cascade, near Durango, CO, 1924. *Photographer: Unknown.*

Power line between Red Mountain and Savage Basin junction
house, Telluride, CO, 1922, elevation 12,500 feet, c. 1920.

*Photographer: P. C. Schools.*

Hauling staves, Cascade, c. 1923. *Photographer: Unknown.*

## David Wood

(known as "The greatest living pioneer of
western Colorado"):

"I was the first freighter in the Uncompahgre Valley in 1877 and
had 500 head of stock, mules, horses, and oxen, trekking across
mountain and valley, on dusty rutty roads, bringing in freight
and hauling out ore. I hauled all the equipment for the new
Telluride Power Co. I was friends with L. L. Nunn, the founder.
My teams were either one, four, six, or eight-horse teams. I
had to have some real mule skinners or drivers to handle so
many head of stock at once. There were two wagons to each
outfit. We hauled many hundreds of thousands of pounds of
heavy machinery to the mines and mills of the San Juans over
routes where there were no roads. I built the 'Dave Wood Road'
between Montrose and Telluride and used it as a shortcut to
that section. Once when the road went out I rebuilt the road
and charged Otto Mears $600 for the job." Note: Mr. Wood is
one of two Civil War veterans alive in Montrose.

*Interview, August 1939*

Power lines between Savage Basin and Camp Bird Mine,
Telluride, CO, c. 1905, elevation 13,200 feet. *Photographer: Unknown.*

Distributing copper cables using mules, Camp Bird Mine, Ouray, CO, 1922. *Photographer: Unknown.*

Hauling forty-foot pipe, Cascade flume, Durango, CO, c. 1923. *Photographer: Unknown.*

## T. E. Peterson,
age ninety-four, former carpenter for WCPC:

"I was a carpenter in charge of the Cascade flume and the Tacoma power flume. I worked during the summer only as it was impossible to work in the winter months. The lumber for both flumes was sawn on Mill Creek near Cascade by a man by the name of Graden."

*Interview done by P. C. Schools, October 1939*

Traffic jam of snow removal equipment, near Cascade Dam, Duranago, CO, 1932. *Photographer: P. C. Schools.*

Twelve hundred horsepower waterwheel, Ames, CO, 1923. *Photographer: P. C. Schools.*

## C. B. Akard,

president of the First National Bank of Montrose, CO:

"I've been a user of electric lights ever since the first light company was organized in 1894 or 1895. The service of the old light companies was exceedingly poor and not reliable. The lights were out half the time and there were many fires due to the fact that the electricians didn't understand wiring. One time a light cord burned through at night at the bank and the bulb dropped to the floor. The night watchman discovered the burning wire, broke the door down, and prevented a serious fire."

*Interview, August 1939*

Hotter's Ranch north of Durango before power lines, 1910. *Photographer: P. C. Schools.*

### H. H. Mendenhall,
one of the first farmers in the Uncompahgre Valley
to have electrical connections in his home:

"The power people required at least five farmers per mile to connect up before it would set poles and install electric fixtures in the farmhouses. I thought it would be impossible to get that many so I called up Al Neale, another old-timer, and we started out to get twenty-five names on our petition for a five-mile stretch of line. We had the whole twenty-five before noon that

day. I called the light company and told them and they told me all the farmers would have their homes wired immediately and the poles would be set and the wire stretched within two weeks. They did this and lights were ready to burn in ten days even before the meters were installed. We had them running for several days—for free—without meters!"

*Interview, August 1939*

**Lester Robuck,**
former WCPC employee:

"I came to the Uncompahgre Valley as a small boy in 1885 and later worked for the 'light company.' I remember the local band used to practice in the boiler room of the power plant."

*Interview, August 1939*

Durango Power Plant, c. 1891. *Photographer: Unknown.*

Ten-foot diameter flume erected at Cascade, Durango, CO, summer 1924. *Photographer: P. C. Schools.*

Constructing Cascade Dam, Durango, CO, 1903. *Photographer: unknown.*

Tacoma Power Plant generators, Durango, CO, July 1910. *Photographer: P. C. Schools.*

Boring holes for tie beams, Telluride, CO, 1915. *Photographer: Unknown.*

## P. C. Schools:

"I remember the breaking of the Trout Lake Dam in 1909. That flood caused about $35,000 in damage to the Ilium Power Plant. The Rio Grande Southern Railway once sued our Telluride Power Co. for $50,000 for damage to their roadbed and the loss of passengers and freight for four months.

"A break in the Tacoma pipeline in 1907 cost the company $40,000. In a windstorm between Red Mountain and Henrietta a half mile of line was blown out of the territory. In 1913 lightning burned out three transformers. Almost every summer losses have been counted from forest fires set by the trains passing through the canyon between Silverton and the Tacoma Plant."

*Interview, September 1939*

Clearing rockslide near Durango, CO, 1937. *Photographer: P. C. Schools.*

P. C. Schools, Electra Lake Dam, Durango, CO, May 1909.
*Photographer: Unknown.*

# PHILIP (P. C.) SCHOOLS:
## An Electrical Engineer with the Eye of An Artist

In the early 1900s, just as cheap, practical electricity is beginning to emerge sporadically here and there in ranch houses, schools, and mines, a man who believes electricity is "the coming thing" enters the scene. He is ahead of his time, having earned a degree in the new field of electrical engineering at a time when most people didn't know what electricity was. He is Philip Schools, called "P. C." by everyone, and he will eventually become the chief engineer for the Western Colorado Power Company. He has a singular talent—he is a gifted photographer, really a photo-historian, with an eye for the drama and beauty in the work it takes to put the power poles into the ground, to string the wire, to clear the aspen groves and put in water flumes, and to install the massive machinery needed to generate power. Most of the photos in this book are his. He hangs nonchalantly from a steel cable, gazing at the camera while sitting on a wooden trolley. His spectacles reflect the icy Animas River below. Although his job with the newly formed Western Colorado Power Company pays the bills, it is his intriguing turn-of-the-century photographs documenting the coming of electricity to rural southwestern Colorado from which his legacy glows.

Schools was born in 1881 just before the first winter snowfall in Crow River, Minnesota, where his father made a living selling firewood to miners before moving the family to Idaho to find better work. In 1890, at the age of nine, Schools's parents sent him to work in the mines.

In those days, this was not unusual for boys his age. But his parents were forward thinking for the times and they insisted that he attend school every morning before work so he wouldn't be a "numskull." And that was unusual.

Schools's parents had to send him to another town to attend the only high school in northern Idaho. There was trouble right away. Schools was a stubborn boy who thought he knew more than the schoolmaster and after only a few days he got expelled.

Camera store photo envelope that belonged to P. C. Schools.

He was furious. He tied up some food in a red bandana and walked for miles to another town where he found a job working sixteen hours a day delivering groceries in a horse-drawn wagon. At night he slept among the store's boxes and packing crates.

Schools worked hard, saved his money, and came up with a good plan for his life. He enrolled in a three-year college preparatory course at the University of Idaho, finishing all the course work in only one year and earning his high school diploma. Although his parents pushed for him to become a mining engineer, he protested that "electricity was the coming thing" (Oral History Collection) and followed his dream. He enrolled in Washington State College in Pullman, Washington, and in 1905

P. C. Schools peering out from behind generator at the Gold King Mine, Ames, CO, February 1909.
*Photographer: Unknown.*

was awarded degrees in both mechanical and electrical engineering.

By 1908, after he had gained experience in other jobs, he was hired by the Tomboy Mine in Telluride as a construction engineer. He did so well there that in 1913 the newly formed Western Colorado Power Company hired him as general superintendent. About this time Schools met and married a beautiful organist at the Methodist Church—so beautiful in fact that he decided then and there to become a Methodist himself.

Schools was in charge of the crews that built power lines over the mountains to the generating plants. He was a hands-on boss, often horseback riding or snowshoeing into the high country to camp out with his crew in a boardinghouse or tent. He always had his camera with him, believing it was important to document every step of the power process. His photographs of powerful looking industrial drums, gears, waterwheels, and swirled complex machinery are beautiful. Equally powerful and sometimes poignant are the photographs of his crews struggling to raise power lines, moving fallen boulders, and swinging hammers. "He never left the house without his camera," remembers his daughter Phyllis, now deceased (Oral History Collection).

Sometimes Schools brought Phyllis to the high country camps with him. She remembered that after saying grace around the dining tent table, the men would raise their long forks in the air ready to jab into the steaming platters of food the cook had prepared. These everyday things, too, Schools photographed.

P. C. Schools on the "car," Durango, CO, 1909. *Photographer: Unknown.*

His photographs have an unexpected quality to them, as in the photo on pages 114–15 of shadowy men taking a smoke break inside a partially built flume.

After more than forty-five years at the power company, Schools retired. He later helped form the Western Colorado Power Company *Old Timers Association* where he enjoyed potluck suppers and reminiscing with his friends.

At the Home for the Aged in Trinidad, Colorado, it is January 1967 and Schools is lying in bed thinking of the days of long ago. His thoughts drift to his boyhood life, to his daughters, and long-dead wife. In his room are yellowing cardboard boxes of his photos. Someday they will be donated to the Center of Southwest Studies in Durango, Colorado, to be enjoyed by researchers, historians, and those who are simply curious about that "coming thing," electricity.

Smoke break, Tacoma flume, Durango, CO, c. 1924. *Photographer: P. C. Schools.*

Building Cascade flume, Durango, CO, 1927. Men used buckets of creosote to seal the wooden flume.

*Photographer: P. C. Schools.*

"I'm thinking over the proposition that you put to me in regard to going to work for you. I am kind of in the dark as to when you would want me . . . I don't know just how long I am going to stay here as Mr. **** has such crazy ideas about erecting that I can hardly stand it. I can't work at a place where I can't do my thing. Jack, I don't want you to let anything influence you in getting a place for me, but the fact that you can use me."

*P. C. Schools, Smuggler, Colorado,*
*September 13, 1908*

P. C. Schools (left), Tacoma Power Plant, Durango, CO, 1949. *Photographer: Unknown.*

Creosoting stubs, Rockwood, 1909. *Photographer: P. C. Schools.*

# "Back in those days one had to be tough— both physically and morally."

Hoadley Maris, head chainman on survey crew, Silverton, CO, c. 1929. *Photographer: Unknown.*

"Back in those days one had to be tough—both physically and morally. All the linemen and patrolmen were deputy sheriffs, and all carried from one to two braces of shooting irons. They never slew anyone—that was never heard of. However, on more than one occasion they had to exert their authority to protect the company's property and honor."

*P. C. Schools, chief engineer*

Linemen raising pole, c. 1930. *Photographer: St. Claire Photo.*

Hoadley Maris, Tacoma-Silverton Line, c. 1933.

*Photographer: P. C. Schools.*

# GLOSSARY

**ALTERNATING CURRENT ("AC"):** Current is electricity in motion. In alternating current the flow of electricity moves forward and backward, switching direction over and over again continuously, 60 times per second. This alternating movement allows for a stronger charge that can travel longer distances and more cheaply than direct current.

**AMES POWER PLANT:** In 1891, in a shack built at the confluence of the Howard's Fork and Lake Fork branches of the roaring San Miguel River, something quite amazing happened. It was here for the first time in the world that alternating current was successfully produced and transmitted, surging 2000 feet and 2.6 miles up the mountain to the Gold King Mine to power it, thus transforming the commercial application of electricity forever.

**BULL GANG:** A crew of men who did heavy manual labor.

**CHAINMAN:** Worked on survey crews holding and positioning chains to take measurements for the placement of power poles.

**COPPER CABLE:** Type of cable used as an electrical conductor in electric wiring since the invention of the electromagnet and telegraph in the 1820s. No.2 copper cable, which is clean, unalloyed copper wiring, was the type of copper cable used most often.

**CREOSOTE:** A tarry substance that was used to weatherize wooden power poles and other wood products.

**CROSSARM:** A wooden arm or bracket that is placed crossways on a power pole to keep wires or cables supported and separated.

**DEAD-END CLAMPS:** Clamps that hold and maintain a wire in a state of tension.

**DIRECT CURRENT (DC):** This type of current starts at one place and flows in only one direction. The DC power system necessitated large, costly distribution wires and forced generating plants to be nearby. In the "War of the Currents" Thomas Edison staked his reputation on DC, telling the world that AC was too dangerous to use.

**ELECTRICAL ENGINEER:** Occupation dealing with the practical application of electricity in all its forms, especially the distribution of power and the design and operation of power company machinery and equipment.

**FLUME:** Man-made channel for water, usually made of wood treated with creosote, to lead water from a dam to a power plant.

**GAUGING STATION:** A site on a river or reservoir where systematic observations of water levels are made so engineers can make predictions about water activity.

**GENERATOR:** A machine that produces electricity using mechanical energy, such as the flow of water stored in dams, to turn turbines. The shaft of the turbine goes into the generator where a series of magnets inside coils of wire are rotated. This process moves electrons, which produces electrical current.

**GOLD KING MINE:** Hanging on at timberline at 12,000 feet elevation above Ophir, CO, the Gold King Mine, discovered by Olaf Nelson in 1887, was going bankrupt. However, under the management of Lucien Lucius Nunn (L. L. Nunn), it was the first mine to be powered by alternating current electricity produced at the Ames Power Plant 2.6 miles below.

**GROUNDMAN:** Worker who digs holes and does the heavy lifting.

Helping the crews build the flume were sure-footed pack mules that lugged beams poles and wire up long, steep, twisting trails. They were affectionately known as "the groundsman's helpers." Silverton-Ouray Line, October 1944. *Photographer: P.C. Schools.*

**HOT TAP:** Energizing a new section of power line by tapping, using an insulated stick, onto an already energized power line.

**HYDROPOWER:** Electricity produced by using water flow to turn turbines.

**JUNCTION HOUSE:** A small structure used to house the termination points of electrical cables.

**KILOWATT ("kW"):** A way of measuring the amount of electrical force that makes electricity move through a wire. A unit of electrical power is called a "watt" and a kilowatt is equal to 1000 watts.

**NUNN, LUCIEN LUCIUS ("L. L.NUNN"):** Born in 1853 in Ohio, this brilliant entrepreneur engineered the world's first successful use of alternating current electricity from the Ames, CO, power plant to the Gold King Mine, changing the course of history and in the process saving the mining industry as well.

**PATROLMAN (LINEMAN):** Installs or repairs cables and wires, and erects power poles and transmission towers.

**SLUICE GATE:** A moveable wood or metal gate that allows water to flow under it. Sluice gates control water levels and flow rates.

**SPILLWAY:** A structure used for the passage of surplus water from a dam into a down-stream area.

**STUB POLES ("STUBS"):** Stubs are set into the ground to assist in holding up long spans of transmission wires from power poles that are on difficult terrain.

**SUBSTATION:** A place where the voltage of electricity is transformed as it passes through on its way from the power plant to homes and businesses.

**SWITCHMAN:** Installs electrical switching equipment, services and installs wiring, and repairs equipment.

**TEAMSTER:** Power plants, dams, and flumes were built by horse and mule power. Teamsters provided teams of these animals plus wagons to do all the hauling, particularly to places inaccessible by train.

**THAW WAGON:** A small, portable generator of heat that was used to thaw ground at construction sites and to "cure" freshly laid concrete.

**TIE BEAMS:** A horizontal beam at the base of a triangular truss that prevents the other two structural parts from separating.

**TRANSFORMER:** Electrical device that transforms electrical energy from one set of voltage to another.

**TROUBLEMAN ("TROUBLE SHOOTER"):** When there is a power problem, this is the person who goes to the site to figure out what the problem is and then fixes it.

**"WAR OF THE CURRENTS":** In the late 1880s, the "War of the Currents" raged between George Westinghouse and Thomas Edison over Edison's promotion of DC power for electric power distribution over AC power, advocated by George Westinghouse. In the end, AC won because it was cheaper and could travel farther than DC.

**WATCHMAN:** Power plants are filled with valuable machinery and materials. The watchman was hired to live on the premises, usually in a small cabin, to guard the power company's assets. Thefts were common in the early days and many employees carried firearms.

**WATERWHEEL (i.e. "PELTON WHEEL"):** A hydropower plant is basically an oversized waterwheel that acts as a turbine, spinning as a stream of water hits its blades, converting the water's energy into electrical energy. A waterwheel consists of a large wooden or metal wheel, with a number of blades or buckets on the outside rim. In the 1870s, Lester Pelton developed the highly efficient Pelton Wheel pictured in many photographs in this book.

**WIREMAN:** Assembles, installs, tests, and maintains electrical wiring equipment.

# SOURCES

Buys, Christian. "Power in the Mountains: Nunn Catapults San Juans into Age of Electricity." *Colorado Heritage Magazine*, Issue 4, 1986.

Cummins, M. L. and P. C. Schools. "An Industry and an Institution of Higher Learning are Born at Ames, Colorado in 1891." *Pioneers of the San Juans*, Vol. IV, pp. 130-34.

Ellison, Todd. Introductory summaries of the WCPC Collection, Center of Southwest Studies, Fort Lewis College, Durango, CO.

Marshall, John B. "The story of J. A. Porter." *Pioneers of the San Juans*, Vol. III, pp. 76-86.

Western Colorado Power Company. Photo Collections P009 &P001, Center of Southwest Studies, Fort Lewis College, Durango, CO.

———."The First 50 Years–A Romance of Electricity and the Western Slope, 1963"; rev. *A Romance of Electricity on the Western Slope.*

———."Western Colorado High Voltage Line Completed Over Hazardous Mountains Without Accident." (WCPC Photo Collection).

———.Personnel records from 1917, and interviews. Center of Southwest Studies, Fort Lewis College, Durango, CO.

Oral History Collection. Letters from P. C. Schools to his daughters, and oral history of Phyllis Schools Case, taken by Todd Ellison, archivist, Center of Southwest Studies, Fort Lewis College, Durango, CO.

## ACKNOWLEDGMENTS . . . Thanks for the Memories

Thank you to the staff at the Center of Southwest Studies in Durango, Colorado, for the freedom to explore and use the amazing special collections, library, and archival materials that they preserve for researchers and students of southwest history. They are, in essence, the stewards of the memories. A special thank-you to Nik Kendziorski who made sure I started off precisely and in the right direction and to Joe Helzer, future historian.

## ABOUT THE AUTHOR

Esther Greenfield was born in Washington, D.C. and grew up in the shadow of the Capitol, enjoying the experiences typical of that city. In fact, she recounts, Harry Truman once tipped his top hat to her during John F. Kennedy's inauguration parade. She is a

retired federal employee whose experience includes a two-year training program sponsored by the White House to train personnel specialists for the Department of Veterans Affairs. Since retiring to Durango, Colorado, she has written many historically oriented articles that have been published in the *Durango Herald* newspaper. In addition, she has collaborated on making, showing, and selling Raku-fired pottery decorated with some of the nearly one thousand historic aspen tree carvings that she has found and recorded while hiking in the aspen groves of the Weminuche Wilderness area near Durango

Esther is a volunteer archivist/researcher at the Center of Southwest Studies, where she digs through old records and photographs, organizing them so that researchers will have ready access. In fact, it was her three-year-long project going through the Western Colorado Power Company Photo Collection at the Center that led directly to this book. The photographs in *Tough Men in Hard Places* represent only a small portion of the approximately eight thousand photographs that she went through. She says the unselected photographs are only slightly less interesting than those that made the final cut for the book.

Esther enjoys hiking with her husband and her dog. And when they walk under the old wooden water flumes, with their lingering tarry smells, she likes to think about the tough men who built them. www.esthergreenfield.com.

CPSIA information can be obtained at www.ICGtesting.com
Printed in the USA
BVOW10s0000121114

374547BV00005BA/8/P